INFORMATION MINING

THE NEW GOLD

By

ALVIN J. SMITH

Copyright © by Alvin J. Smith 2022. All rights are reserved.

Before the document is duplicated or reproduced in any manner, the publisher's consent must be gained.

Therefore, the contents within can neither be stored electronically, transferred, nor kept in a database.

Neither in part can the document be copied, scanned, faxed, or retained without approval from the publisher or creator.

Table of contents

Introduction	4
Chapter one	6
What is Information Mining?	6
Chapter Two	14
Information mining methods	14
Chapter three	21
Information mining devices	21
Chapter four	31
Information mining virtual products	31
Conclusion	41

Introduction

Enormous information driven associations instill a gigantic measure of information. Today, most information exists in crude structure inside these associations. In addition, associations have been utilizing procedures to deal with information and finding business bits of knowledge through it for a long time.
Notwithstanding, for associations gathering a mammoth measure of information, a few hearty information handling procedures like Information digging for information disclosure of information (IDI) and it are an unquestionable requirement to deal with information.

With the headway of manageable assembling advances and the improvement of the world's assembling industry, fabricating organizations are giving increasingly more consideration to the association among items and assembling information. Through the data overhaul of assembling hardware and programming, information in the area fields can be gained, put away, and dissected all the while in the assembling system and afterward is taken care of back to creation to further develop the creation effectiveness and yield, to abbreviate the assembling cycle, and to further develop item quality. This has turned into a pattern in the assembling business. Verifiable information, like plan data and assembling data in an undertaking, contain a rich item plan and assembling information. Mining and examination have turned into a significant method for undertakings to improve their seriousness. In the assembling business, process arranging is a sort of involvement based complex information application action. The mining of cycle arranging information can guarantee item quality. Consequently, deciding how to transform the information into valuable information that helps fabricating process diagnostics and further develops item quality has turned into the focal point of examination

Chapter one

What is Information Mining?

Information mining alludes to a cycle including the modification and extraction of valuable information examples, patterns, and connections from colossal crude informational indexes. Information mining assists with keeping a knowledge over information examination

Besides, information digging includes various calculations for changing crude information in an important structure. Likewise, a vigorous strategy for handling information to improve decision-production for associations through the inclusion of natural examination of information with the assistance of AI.

Also, information mining procedures are fit for arranging, separating, changing, and springing up important bits of knowledge inside huge informational indexes. Associations gather a mammoth measure of information in crude configuration, information mining carries the most pivotal data to the front.

Information Mining Interaction

Regardless, the course of information mining is bipartite. The two sections incorporate information preprocessing and information mining.

Information Preprocessing

The information preprocessing stage incorporates purging, joining, modifying, and changing information. Interestingly, the information preprocessing stage is pivotal for totally outfitting the information to drive accomplishment for information driven associations.

Likewise, a few elements hamper the efficiency of the interaction. These incorporate; information accuracy, information steadiness, information rightness, information culmination, and information conveyance time. Plus, performing examinations on and off base and conflicting information can never give precise outcomes. Consequently, there are major areas of strength for information preprocessing. Information preprocessing stages incorporate;

Information Cleaning

The primary stage for preprocessing the information incorporates cleaning the informational collection. For certain, information cleaning is urgent for setting up the crude type of information through the evacuation of pointless and tedious information, filling in the missing qualities, and coordinating the crude information.

Information Coordination
The handled and clean information is as yet having a place with different sources like data sets, information distribution centers, and so forth. Information Joining includes uniting valuable and clean information for examination to accelerate the information mining process.

Information Decrease
Information decrease is tied in with picking relevant information from the informational collection or information source. One of the ways of playing out this interaction is through utilizing a brain organization. Moreover, different methodologies for performing information decrease like dimensionality decrease, numerosity decrease, and information pressure.

Information Change
The change stage includes changing the information into an allowable and economical configuration for further developing the information mining process. Information change procedure incorporates code creating interaction and information planning.

Information Mining

The information mining stage includes information mining, evaluating various examples inside handled informational indexes, lastly addressing experiences or information for the information.

Information Mining

Information mining is involved by huge information associations for sorting out business drifts and producing business knowledge. Be that as it may, information mining is simply trustworthy upon information assortment and preprocessing for precision.

During information mining, some business factors are set to decide examples and connections inside the information involving models for arrangement and strategies for information bunching.

Assessment of Examples

This stage works after using the handled and clean information completely. It goes through different examples. Particularly, while working with AI, brain organizations, and GPU, developing various convincing examples continuously.

Certain associations are worked inside the information to totally comprehend it through utilizing techniques like Information representation and Information synopsis.

Portrayal of Information

The information assembled and the information mined should be addressed in machine-reasonable organization. During this stage, we use apparatuses like information perception and information portrayal.

The Ascent of Information Mining in Security and How It Affects What's to come

The opportunities for new or extended open doors are just about boundless, re-imagining "information security" and not that far in the distance, perhaps the way in which you depict your business.

What business would you say you are ready for? I'm betting you said, "Security," or maybe "Alerts," or perhaps "Frameworks coordination." And those intensely centered around observing or different administrations might say, "Repeating income." Those are just fine and appear to be legit today.

Yet, because of propelling innovation and advancing client needs/interests, tremendous changes have arrived at this industry the recent many years that incited rethinking reason and portrayal (for example Public Criminal and Alarm Relationship to Electronic Security Affiliation and Focal Station Caution Relationship to The Checking Affiliation). A change keeps on unfurling.

"The business is simply starting to understand that information is our new money," It is being utilized to make clients safer and more viable in anything they want to achieve. The present innovation permits us to convey clients a more powerful encounter than any other time. The key is giving a dependable, secure arrangement that gives them the security and comfort they need and need. Secure, helpful, private, it's all fundamental."

After sixty years, electronic security is among numerous enterprises being influenced by another social shift. Driven primarily by more youthful grown-ups weaned on Web and cell phone empowered moment delight. Yet additionally penetrating all rungs of current American culture — in the present everything on-request world clients and buyers look for hotter innovation, way of life improvement and accommodation.

The ramifications of this shift for our industry are huge all through the channel, influencing how arrangements are planned and showcased as well as how administrations are created and conveyed. New difficulties proliferate yet the open door is irrefutably sound, and the potential result is endless.

As a security vendor or integrator, take stock to guarantee your business' way of life, cycles and contributions are where the present clients are … for it's there that achievement is standing by.

Where is information mining (IM) utilized?

It is generally utilized in different ventures, including medical services, retail, money, government, and assembling.

For instance, if an organization has any desire to find examples or patterns among clients that purchase specific items, it can utilize information mining strategies to investigate their buying history and foster models that foresee which clients wish to purchase explicit products in light of their socio economics or conduct. Along these lines, in retail, information mining assists organizations with growing more fruitful deals procedures.

What's more, these instruments can be utilized to:

Portion clients: recognize gatherings of clients that share comparable ways of behaving and target them with customized promoting messages.

Anticipate cancellations: figure out which clients will quite often drop their orders in view of authentic information.

Distinguish misrepresentation: in light of verifiable exchange information, recognizing dubious ways of behaving and blocking it is conceivable.

Prescribe items and administrations to clients relying upon their previous experience

Models in different regions

Information mining methods are additionally making strides in training, science, coordinated factors, money, and banking - as such, basically every circle.

In schooling, IM helps assemble redid programs in view of:

- Understudies' learning designs - for example, their propensities to consume data through video sound, text, or a blend of the three.
- Work market patterns - this permits to decide the most important instructive concentration.

In finance, information mining is utilized to:

- distinguish venture open doors;
- anticipate interest for a few stock offers, which empowers expected financial backers to settle on informed choices.
- Information mining likewise has applications in policing knowledge:

Customs officials can more readily comprehend the average profile of boundary violations in light of line crossing history and spotlight on unambiguous classifications of people.

Police can distinguish regions where they need to send more labor, knowing when and where the probability of wrongdoing is most noteworthy.

Chapter Two

Information mining methods

To extricate data from information, a wide assortment of information mining methods are utilized.

These include:
- order
- grouping
- affiliation rule learning
- relapse
- inconsistency discovery
- successive example mining

Contingent upon information qualities, clump or constant handling can be utilized. The first works for large measures of information gathered over a specific period. Constant handling applies to frameworks with progressively refreshed information, the case of which is Google Examination's ongoing outline report that mirrors the site client movement occurring at this very moment.

Characterization
Arrangement is utilized to separate information into foreordained gatherings or classes. This information mining strategy decides the class to which a record has a place in view of the upsides of a few credits. The objective is to sort information into predefined classes.

Most usually, grouping includes foreseeing an objective variable that can take on one of at least two potential qualities (e.g., spam/not spam; positive or unbiased/negative survey) given at least one information factor called indicators.

Grouping
Bunching is a strategy for gathering related passages in a data set into groups in light of their similarities. While characterization doles out factors into known classifications, the bunching procedure first singles out these groups in the dataset and afterward bunches factors in light of their qualities.

For instance, you can bunch clients into bunches as per the deals information - the people who routinely purchase pet food or explicit beverages and who are steady in their inclinations and client conduct. When you lay out these groups, you can undoubtedly target them with tweaked promotions.

Bunching has a great many applications:
- clinical diagnostics
- computational science
- text mining
- web investigation
- Affiliation rule learning

Affiliation rule learning finds on the off chance that examples between at least two factors. The least complex model is the relationship between purchasing bread and butter. Individuals that purchase bread generally get margarine with it, as well as the other way around. To that end you will find these two items near each other in a supermarket.

In any case, the connection might not be unreasonably immediate. For example, in 2004 Walmart found that the deals of Strawberry Pop-Tarts were at their top before the tropical storm. Individuals loaded up the necessities like batteries as well as these famous pastries. Everything considered, the mental inspiration is very self-evident: during crises, your number one food provides you with a conviction that all is good, and tarts with a long time span of usability are an ideal choice. In any case, to decide this relationship, it was important to apply information mining methods.

Relapse
Relapse lays out a connection between factors. Objective is to find the right capability that depicts the relationship. In the event that a direct capability (y = hatchet + b) is utilized, the cycle is called straight relapse examination. For different kinds of conditions, techniques like various direct relapses, polynomial relapses, and so on can be utilized.

Straightforward direct relapse
Straightforward direct relapse utilizes one free factor to make sense of or foresee the result.

For instance, you have a table with the example information concerning the temperature of links and their strength. Presently, you can do straightforward direct relapse to make a model that can foresee the solidness of a link in view of its temperature.

temperature solidness table

The forecasts you make with basic relapse will generally be somewhat mistaken. A link's toughness relies upon numerous different things than simply the temperature: wear, weight of carriage, stickiness, and different elements. To that end basic direct relapse isn't typically used to settle genuine errands.

Numerous direct relapse for AI

Not at all like basic direct relapse, numerous straight relapses utilize a few informative factors to foresee the reliant result of a reaction variable.

A numerous direct relapse model seems to be this:

$$Y = a + b\_1X\_1 + b\_2X\_2 + b\_3X\_3 + ... + b\_tX\_t$$

Here, YY is the variable that you are attempting to foresee, XX's are the factors that you are utilizing to anticipate YY, aa is the catch, and bb's are the relapse coefficients - they show how much an adjustment of specific XX predicts an adjustment of YY, all the other things being equivalent.

In actuality, different relapses can be utilized by ML-fueled calculations to anticipate the cost of stocks in light of fluctuations in comparable stocks.

Nonetheless, it would be incorrect to say that the more factors you have, the more exact your ML forecast is.

Issues with different direct relapse

Two potential issues emerge with the utilization of different relapses: overfitting and multicollinearity.

Overfitting implies that the model you work with different relapses turns out to be excessively restricted and doesn't sum up well. It works alright on the preparation set of your AI model however doesn't work as expected on the things not referenced previously.

Multicollinearity portrays what is going on when there is a relationship between isn't just the free factors and the reliant variable yet additionally between the autonomous factors themselves. We don't believe that this should happen on the grounds that it prompts deceiving results for the model.

Its most considered normal application is arranging and displaying. One model is anticipating clients' age in light of their buy history. We can likewise foresee costs in view of such factors as customer interest - for instance, a flood of costs on the optional market because of the expanded interest for vehicles in the US.

Abnormality discovery

Irregularity discovery is an information mining method used to distinguish exceptions (esteems that veer off from the standard). For instance, in online business datasets, it can distinguish uncommon deals during a given week at a store area. In addition to other things, it very well may be utilized to find credit or charge extortion and recognize interruption or break in the organization.

Consecutive example mining
Consecutive example mining is an information mining region that identifies significant connections between events. Recognizing a period requested succession of occasions that occur with a particular recurrence permits us to discuss a reliance between them.

Consecutive example mining is an information digging technique for getting regular successive examples in a successive data set. The above specialists read up various calculations for the assembling system to enhance the activity grouping and utilized a consecutive example mining calculation to mine various sorts of information. In any case, they didn't join their strategies for looking for the principles behind assembling highlights according to the viewpoint of finding quality reasons. A novel, regularly shut successive example mining calculation in view of the assembling system of text context oriented semantics and the assembling system unit was proposed to get key variables of item quality.

Suppose we need to research the effect of a drug or a specific restorative technique on the future of disease patients. Consecutive example mining empowers you to do that by adding a worldly aspect to the investigation. This method is material, among others, in medication to work out the request for a patient's clinical solutions and in network protection to anticipate potential assaults on the framework.

Uses of successive example mining include:

- shopping groupings
- financial exchanges
- cataclysmic events
- clinical medicines
- DNA sequencing research

Chapter three

Information mining devices

Information mining devices are programming arrangements that utilize man-made intelligence and ML to pull and break down information, feature drifts, and give noteworthy experiences to organizations. The product can refine data from both organized and unstructured datasets, so associations can make forecasts and grasp connections between various pieces of their business. Information mining apparatuses permit organizations to resolve questions that would require some investment to reply assuming they needed to examine information manually.

The accompanying information mining apparatuses all have great client audits and solid capabilities.

RapidMiner
RapidMiner offers computerized information mining and demonstrating instruments with artificial intelligence and ML to give clear representations and prescient investigation. The simplified connection point makes it simpler for investigators to make prescient models, and the library incorporates more than 1,500 prefabricated calculations, important because there's a model for almost any utilization case. There are additionally pre-constructed layouts for normal situations, including misrepresentation discovery and upkeep, to bring down the time experts need to spend building the models. RapidMiner can interface with any information source, or clients can import information from Succeed. Closely involved individuals should demand estimating from RapidMiner; it's not accessible on the site.

21

Key Highlights
- Point-and-snap data set associations
- Intuitive model developer
- MySQL, PostgreSQL, and Google BigQuery support
- Out-of-the-container calculations and layouts
- Various sorts of diagrams and charts
- Mechanized AI
- R and Python support

Experts
- Direct associations with outside information sources
- Simple to mechanize whole AI process
- Supportive and responsive client assistance

Cons
- A few clients said the web application for simulated intelligence Center doesn't have a lot of usefulness
- The expense is higher contrasted with contender stages

Prophet Information Digger
Prophet Information Digger is an expansion of the Prophet SQL Designer that assists analysts with rapidly constructing an assortment of AI models, applying them to new information, and looking at the models for noteworthy experiences. It offers an intuitive proofreader, permitting the two information researchers and standard clients to find solutions to their information related questions. The work process Programming interface makes it simpler to convey the model all

through the business, implanting investigation into the applications where investigators are as of now working. Estimating isn't plainly accessible on the site, so organizations should contact Prophet for more data.

Key Highlights
- Simplified model developer
- Intuitive work process device
- Different kinds of representations
- Joining with open-source R
- Robotized model structure
- Works with BigDataSQL to get to significant information sources

Geniuses
- Can ingest both organized and unstructured information
- Simple to get and rebuild information
- Stage is coordinated and gives simple information the executives

Cons
- The point of interaction may not be basically as easy to use as different stages
- A few clients griped the handling was slow

Sisense
Sisense is information examination programming that permits clients to implant examination into the stages they as of now work in, placing the data in a similar spot they're deciding. Moreover, organizations can white name the inserted examination, so they can likewise

push them out to their clients. With live information associations, organizations can get ongoing bits of knowledge and a solid self-administration stage. With code-first, low-code, and no-code choices accessible, examiners of any expertise level can get their information questions addressed and assemble accommodating models. Besides, the artificial intelligence permits examiners to type in an inquiry, and afterward it guides them through the examination. Evaluation isn't accessible on the site.

Key Elements:
- Prescient examination
- Code-first, low-code, and no-code instruments
- Self-administration examination
- Live information associations
- Implanted examination
- Cloud-based choices

Geniuses
- Gives profound bits of knowledge into information
- Simple to utilize and make dashboards and inquiries
- Rapidly associates with data sets and cycles information

Cons
- Doesn't necessarily in every case save questions clients are dealing with
- Reports don't necessarily in every case update in the timetable clients set

Alteryx APA

Alteryx APA offers mechanized investigation with AI across the whole cycle, including mining, demonstrating, and perception. There are north of 80 locally coordinated information sources that clients can pull from, including Prophet, Amazon, and Salesforce, or they can utilize APIs to interface with others. Examiners can likewise add guides to their perceptions to feature geographic patterns. Alteryx offers bit by bit advisers for assisting examiners of any ability with evening out form models without coding. Notwithstanding, master information experts can likewise utilize R-based models. Estimating data isn't accessible on the site.

Key Elements
- Mechanized examination
- Local information source mixes and APIs
- Geographic examination
- No-code choices
- Various representation choices
- Sharing and sending out abilities

Masters
- Solid and productive foundation
- Upholds cycles of all sizes and levels of intricacy
- More easy to use than comparable stages

Cons
- Large information sources once in a while consume most of the day to process
- Does exclude however many visual instruments as contenders

SAS Information Mining

SAS Information Mining assists associations answer complex inquiries with investigation through robotized demonstration and a cooperative stage. With normal language age, the stage can make a post-project outline, enumerating significant patterns, exceptions, and experiences. Then, at that point, clients can add notes to the report to make correspondence and joint effort simpler. SAS Information Mining upholds an assortment of coding choices, so examiners can make or change calculations in their language of decision. Information researchers can likewise consolidate organized and unstructured information in models to get however much data as could be expected. Evaluation isn't accessible on the SAS site.

Key Elements
- Intuitive connection point
- Code-first and no-code choices accessible
- PDF sharing
- Cooperative climate
- Public Programming interface
- Programmed demonstrating
- Regular language handling

Experts
- Accommodating and responsive client assistance
- Simple to incorporate information
- Huge number of calculations accessible

Cons

- A few clients grumbled that the stage wasn't refreshed all the time
- Hard to decide best practices for the instrument

Teradata

Teradata is an information digging device used for associations utilizing multi-cloud organizations, giving admittance to all data sets, information lakes, and outside SaaS applications. No-code choices permit clients from any business division to find solutions to their inquiries to pursue more educated choices. Associations can send Teradata on any of the significant public cloud stages, including AWS, Sky blue, and Google, as well as in confidential mists or on-premises. Teradata doesn't charge forthright expenses, rather offering a pay-more only as costs arise model. A valuable mini-computer is accessible on the site to assist clients with assessing their expenses.

Key Highlights
- Code-first and no-code choices
- Adaptable responsibilities
- Numerous organization choices
- Incorporates with different sources
- Support for all normal information types and organizations
- Job based examination choices

Experts
- Solidifies information from all sources
- Handles refined and straightforward inquiries

- Requires almost no upkeep for the cloud-based choices

Cons
- Can be costly contrasted with contender stages
- On-premises upkeep can be troublesome and tedious

Dundas BI

Dundas BI is an information investigation stage that offers continuous experiences and outwardly engaging reports and dashboards. It can unite information from any source with open APIs, guaranteeing that clients have all the data they need to make successful models. Clients can make content that is straightforward with negligible contribution from IT. Intuitive dashboards permit investigators to alter models to perceive what various factors would mean for the business. Dundas BI offers a ton of out-of-the-crate usefulness without requiring additional items or overhauls. Valuing data isn't accessible on the site.

Key Highlights
- Adaptable dashboards
- Open APIs
- Simplified plan devices
- Different representation choices
- Correspondence and joint effort apparatuses
- Robotized notices
- What-if investigation

Masters
- Highlight rich stage

- Seriously estimated contrasted with comparative stages
- Functions admirably on cell phones and work areas

Cons
- Can have a lofty expectation to learn and adapt
- A few clients whined about the stage crashing

H2O

H2O is a simulated intelligence cloud used for information mining to further develop the bits of knowledge organizations get from their information and their independent direction. Robotized AI takes care of mind boggling issues while giving outcomes in a straightforward arrangement. Examiners can prepare and send the simulated intelligence in any climate, and there are a few different demonstrating types that they can browse. Continuous information investigation gives exact forecasts and quick bits of knowledge to assist organizations with settling on faster choices and work on their adaptability. H2O can be conveyed with one or the other mixture or completely oversaw choices. The stage is open-source and allowed to utilize, yet organizations can pay for big business backing and the board.

Key Highlights
- Open-source stage
- Strong artificial intelligence calculations
- Support for various programming dialects, including R and Python
- Programmed tuning and preparing of ML

- In-memory handling
- Simple sending

Geniuses
- Works on model precision and execution
- Simple to get and figure out how to utilize
- Gives active instructing

Cons
- A few clients need more granular control
- No help for edge processing

Information Mining Prompts Noteworthy Experiences

Organizations that utilize information mining programming get quicker admittance to significant data and noteworthy bits of knowledge that can further develop their dynamic interaction. Every day, organizations take in such an excess of information that it would be difficult to figure out physically. They need information mining devices that incorporate man-made intelligence to run, consider the possibility of situations and get exact figures. Organizations searching for the best information digging programming for their business ought to exploit free preliminaries and read client surveys to figure out which one will turn out best for their group

Chapter four

Information mining virtual products

Hexomatic

Hexomatic is a no-code, work computerization stage that empowers clients to use the web as their own information source, influencing instant mechanization to scale tedious errands. Scratch any site, track down designated leads, and advance information in minutes.

With this stage, there is no coding, no recruiting, just Lego-style blocks you can assemble to make your ideal work process to mechanize redundant undertakings at scale, so your group can zero in on higher-esteem work.

Ordinary clients
- Consultants
- Independent companies

- Medium size organizations
- Huge endeavors

Most esteemed highlights by clients
- Alarms/Warnings
- Action Dashboard
- Programming interface
- Work process The board
- Information Import/Commodity
- No-Code
- Configurable Work process
- Business Interaction Computerization

IntelliFront BI

IntelliFront BI is an Information Examination and Business Knowledge arrangement intended for the basic requirements of organizations, assisting with conveying data and understanding to representatives, accomplices, and clients. IntelliFront BI gives answers for detailing, circulation, dashboards, KPIs, planning, and work processes.

Key advantages of utilizing IntelliFront BI
- Plan and serve outwardly shocking, intuitive, continuous reports and dashboards in the program based administrator module utilizing a wide exhibit of visuals.
- Add naturally refreshing and intuitive KPI cards with total, adjusting, estimation units, objectives, colors and glyphs.
- Give clients a straightforward, secure and natural detailing gateway to consume their Reports,

- KPIs and Dashboards as a feature of their day to day schedules.
- Secure your information with Dynamic Registry Joining, Single Sign On and 2-Element Validation as standard.
- Set up client bunches with both Promotion and non-Promotion clients, permitting you to safely convey reports and intelligent BI dashboards inside and outside your association with no extra authorizing costs.

Mix with ChristianSteven's planning suite permits you to computerize the product and conveyance of reports in standard configurations like PDF, CSV and XML to various objections like Email, Printer, FAX, Envelope, Google Sheets, Google Drive, Sharepoint Dropbox, Slack and more.

Average clients
- Independent companies
- Medium size organizations
- Huge ventures

Most esteemed highlights by clients
- Information Representation
- Announcing/Investigation
- Information Import/Product
- Visual Examination
- Programming interface
- Announcing and Insights
- Dashboard

- Search/Channel

SCALUE

SCALUE is a business-execution board programming that helps obtainment groups gain constant knowledge into complete spending to find stowed expenses and failures across business processes. Heads can arrange unstructured acquisition information and screen key execution pointers (KEPs).

The stage empowers supervisors to characterize project targets and representative obligations among staff individuals on a bound together connection point. SCALUE additionally permits administrators to identify bottlenecks across acquirement cycles and track compelling benefit and misfortune (P&L) reserve funds.

Ordinary clients
- Independent companies
- Medium size organizations
- Huge endeavors

Most esteemed highlights by clients
- Programming interface
- Access Controls/Consents
- Detailing/Investigation
- Outsider Incorporations

- Movement Dashboard
- Cautions/Notices
- Revealing and Measurements
- Observing

FSWorks

FSWorks is an electronic quality control arrangement, which helps organizations in ventures like food and refreshment, car, clinical, and fabricating upgrade creation processes. Key highlights incorporate continuous status following, work directions, pattern investigation, and information securing.

The application empowers clients to quantify the advancement of creation lines through different cycle measurements like standard deviation, Cp, Cpk, and that's just the beginning. It assists clients with creating graphical portrayals to see creation line execution all through a characterized length and distinguish gambles/limits related with the cycle. Chiefs can utilize ANDON presentations to picture machine, line, or cycle status, giving clients recognizing shortcomings accessing the framework.

FSWorks incorporates Generally Gear Viability (GGV) shows, which permit workers to assess and examine machine execution and quality. The arrangement likewise empowers managers to transfer design arrangements, change useful cutoff points, and alter data for administrators. Furthermore, clients can record and keep up with occasional logs of cautions, changes, and updates in the framework, alongside remarks added by administrators.

Common clients
- Independent ventures
- Fair size organizations
- Huge ventures

Most esteemed highlights by clients
- Action Dashboard
- Announcing and Insights
- Announcing/Investigation
- Adjustable Reports
- Work process The board
- Information Import/Commodity
- Constant Information
- Movement Following

Octoparse
Octoparse is a product for programmed information extraction. By utilizing this device, clients can scratch web information without coding and transform pages into organized information documents inside a few ticks. So, the easy to understand apparatus permits clients to get web information without the requirement for complex programming information.

With the implicit formats, Octoparse empowers clients to begin web scratching from numerous well known sites in seconds without arranging the undertaking without help from anyone else.

Other valuable highlights incorporate programmed recognition, Programming interface access, planned execution, cloud-based execution, IP revolution, and so forth.

Run of the mill clients
- Consultants
- Independent companies
- Medium size organizations
- Huge endeavors

Most esteemed highlights by clients
- Action Dashboard
- Programming interface
- Information Import/Product
- Continuous Information
- Work process The board
- Web Information Extraction
- IP Revolution
- Configurable Work process

Data Cadabra

data cadabra is an information science answer for promoting and CRM groups. It can go with clients all through the information abuse chain (information arrangement, client conduct examination, enactment, and focusing) to empower them to work on the presentation of client activity activities. datacadabra empowers showcasing and CRM groups to utilize client information all the more successfully to further develop their client information and the exhibition of their client liveliness technique.

Key advantages of utilizing data cadabra
As a genuine virtual information researcher, datacadabra enjoys the accompanying benefits:

- Expanded independence for groups who will actually want to set up complex calculations in a natural manner
- straightforwardness on account of bundled calculations that make the execution of investigations less complicated
- efficient because of the improved on investigation process
- functional execution gains because of the nature of the calculations carried out and their effect on focusing on
- examinations remarked naturally because of reports

This large number of components make data cadabra an imaginative arrangement that permits promoting groups to surpass their goals.

Run of the mill clients
- Independent companies
- Medium size organizations
- Huge undertakings

Most esteemed highlights by clients
- Announcing/Examination
- Programming interface
- Announcing and Measurements

- Action Dashboard
- Information Import/Product
- Crusade Examination
- Pattern Examination
- No-Code

Lucidworks Combination

Lucidworks Combination is a cloud-based information disclosure stage, which assists undertakings with performing mental pursuit and producing customized bits of knowledge and proactive proposals. Highlights incorporate validation, visual utilization examination, information grouping, adaptable dashboard, and A/B testing.

The stage uses AI innovation during information ingestion to arrange each datum set and concentrate data from discourse, text, or unstructured information utilizing normal language handling (NLP). Based on Apache Solr and Flash, it records and stores information for continuous disclosure, directs a full-text search utilizing SQL, and processes various questions. Lucidworks Combination predicts client plan and conduct, producing hyper-customized results for the crowd.

It incorporates an application studio, which permits information researchers to make and send information revelation applications for web or portable. The information diagram deciphers question content and permits groups to divide client created data between gatherings. Designers can likewise use a few particular parts like information representation, geospatial planning, route, and then some

Run of the mill clients
- Independent companies
- Medium size organizations
- Huge ventures

Most esteemed highlights by clients
- Information Representation
- Announcing/Investigation
- Visual Investigation
- Search/Channel
- Action Dashboard
- Prescient Examination
- Portable Access
- Regular Language Search

Conclusion

By and large, information mining is viewed as one of the fundamental pieces of AI. Since it is utilized to figure out significant secret patterns and examples inside tremendous volumes of information, both AI and information mining utilize current calculations to uncover related information designs.

With the utilization of an expansive scope of cutting edge methods, one can undoubtedly utilize this result to reduce expenses, lessen gambles, increment incomes, further develop client connections, and so forth.

www.ingramcontent.com/pod-product-compliance
Lightning Source LLC
Chambersburg PA
CBHW050320220526
45465CB00005B/2069